MATH MINDERS

Graphing
GRADE 3

Written by
Vicky Shiotsu

Illustrated by
Becky Radtke

Cover Illustration
by Susan Cumbow

FS112037 Graphing Grade 3
All rights reserved—Printed in the U.S.A.

Copyright © 1999 Frank Schaffer Publications, Inc.
23740 Hawthorne Blvd.
Torrance, CA 90505

TABLE OF CONTENTS

Introduction	2
At the Circus (Reading a picture graph)	3
A Fruity Graph (Reading a vertical bar graph)	4
Let's Recycle (Reading a horizontal bar graph)	5
Crafty Items (Solving word problems/Picture graph)	6
Museum Visitors (Solving word problems/Bar graph)	7
Shopping Fun (Reading a circle graph)	8
All in the Family (Comparing parts of a whole/Circle graph)	9
How Many People? (Comparing populations/Picture graph)	10
Dinosaur Days (Comparing heights/Bar graph)	11
Long and Slippery (Comparing lengths/Bar graph)	12
Changes Over Time (Reading a line graph)	13
Lemonade Sales (Solving word problems/Line graph)	14
An Insect Hunt (Making computations/Completing a picture graph)	15
Sports Day Fun (Making computations/Completing a bar graph)	16
A Stamp Collection (Logical thinking/Completing a picture graph)	17
T-Shirt Count (Logical thinking/Completing a circle graph)	18
Tasty Sandwiches (Using data from a chart to make a picture graph)	19
How Speedy? (Using data from a chart to make a bar graph)	20
All Kinds of Books (Using data from a chart to make a circle graph)	21
Oletown's Rain (Using data from a chart to make a bar graph)	22
Aquarium Animals (Reading tally marks/Creating a bar graph)	23
A Choice of Graphs (Displaying information on two different graphs)	24
Carnival Time (Gathering data from two graphs)	25
A Penny Drive (Gathering data from two graphs)	26
Sizes of Families (Collecting data/Creating a picture graph)	27
Off to School (Collecting data/Creating a bar graph)	28
Coin Throw (Collecting data/Creating a circle graph)	29
TV Watch (Collecting data/Creating a line graph)	30
Answers	31–32

Notice! Student pages may be reproduced by the classroom teacher for classroom use only, not for commercial resale. No part of this publication may be reproduced for storage in a retrieval system, or transmitted in any form or by any means—electronic, mechanical, recording, etc.—without the prior written permission of the publisher. Reproduction of these materials for an entire school or school system is strictly prohibited.

INTRODUCTION

This book has been designed to help students succeed in math. It is part of the *Math Minders* series that provides students with opportunities to practice math skills that they will use throughout their lives.

The activities in this book have been created to help students feel confident about working with graphs. The beginning pages review concepts introduced in earlier grades; the pages gradually increase in difficulty as students learn new concepts and acquire more sophisticated graphing skills. Vocabulary is kept at a third-grade level to help ensure student success.

Various formats are used throughout the book to help maintain student interest. Students will be able to progress at their own speed, using the skill learned in one activity to advance to the next level of understanding. The concepts presented in this book can be taught in the classroom or at home. Students will practice a variety of skills, such as reading different kinds of graphs (picture graph, bar graph, circle graph, line graph), collecting data, and recording information on graphs.

Graphing
GRADE 3

Name_____

At the Circus
Reading a picture graph

The Super Duper Circus has many wonderful performers. The picture graph below shows the number of performers.

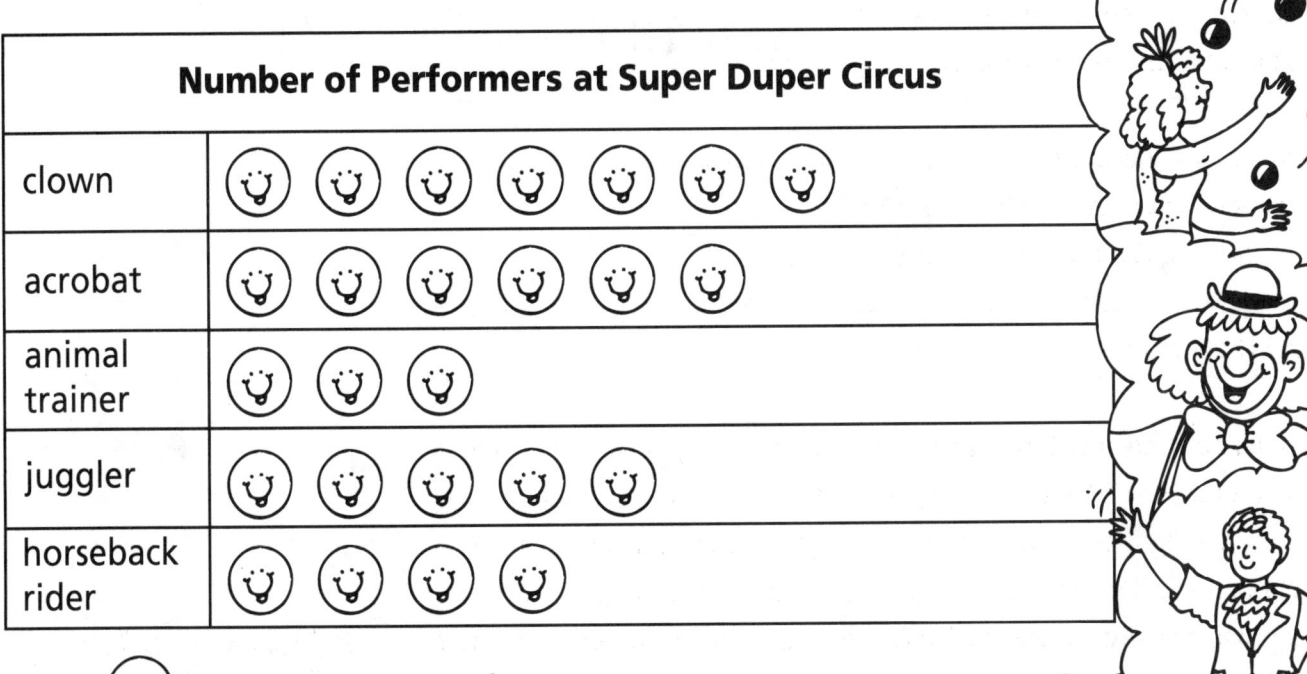

Number of Performers at Super Duper Circus

clown	☺ ☺ ☺ ☺ ☺ ☺ ☺
acrobat	☺ ☺ ☺ ☺ ☺ ☺
animal trainer	☺ ☺ ☺
juggler	☺ ☺ ☺ ☺ ☺
horseback rider	☺ ☺ ☺ ☺

Each ☺ stands for one performer.

Use the picture graph to answer the questions.

A. How many clowns work at the circus? _____

B. Are there more jugglers or acrobats? _____

C. How many animal trainers are there? _____

D. How many more clowns are there than horseback riders? _____

E. Which group has the most performers? _____

F. Which group has the fewest performers? _____

G. What is the total number of performers shown on the graph? _____

© Frank Schaffer Publications, Inc.

Graphing Grade 3

Name_____

A Fruity Graph

Reading a vertical bar graph

Mr. Ortega's students took a survey in class to see which fruits they like the best. The results of the survey are shown on the bar graph below.

Favorite Fruits

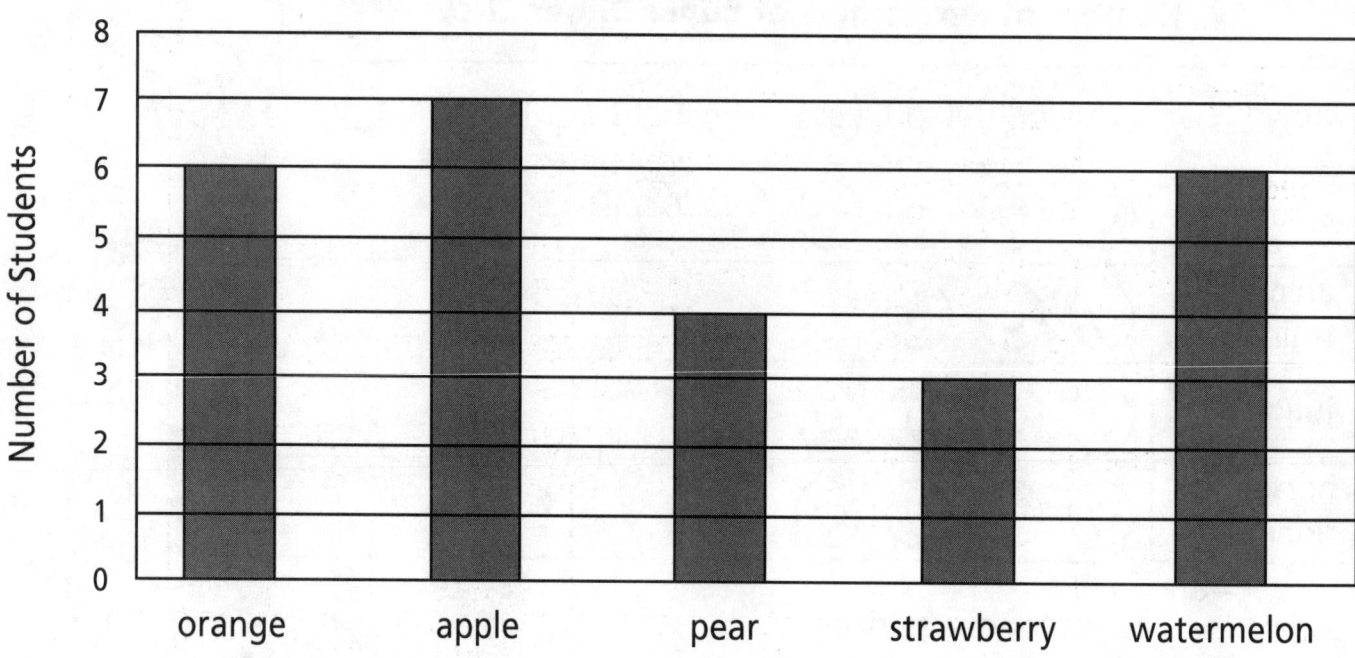

(Bar graph: Number of Students vs Fruit)
- orange: 6
- apple: 7
- pear: 4
- strawberry: 3
- watermelon: 6

A. How many students like pears the best? _____

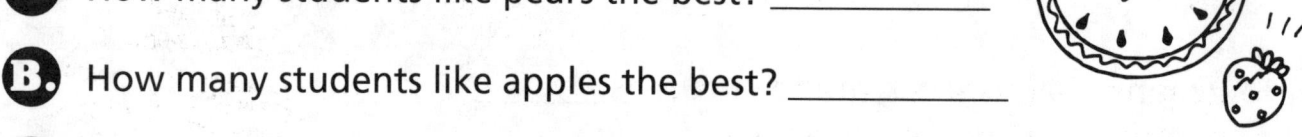

B. How many students like apples the best? _____

C. How many students like strawberries the best? _____

D. Which did more students like—apples or oranges? _____

E. Which did more students like—oranges or pears? _____

F. Which are more popular—pears or watermelons? _____

G. Which fruit is the most popular in Mr. Ortega's class? _____

H. How many students took the survey? _____

Name_____

Let's Recycle............................ Reading a horizontal bar graph

The students at Green Valley School collected aluminum cans for recycling. Look at the bar graph and answer the questions.

Number of Cans Collected

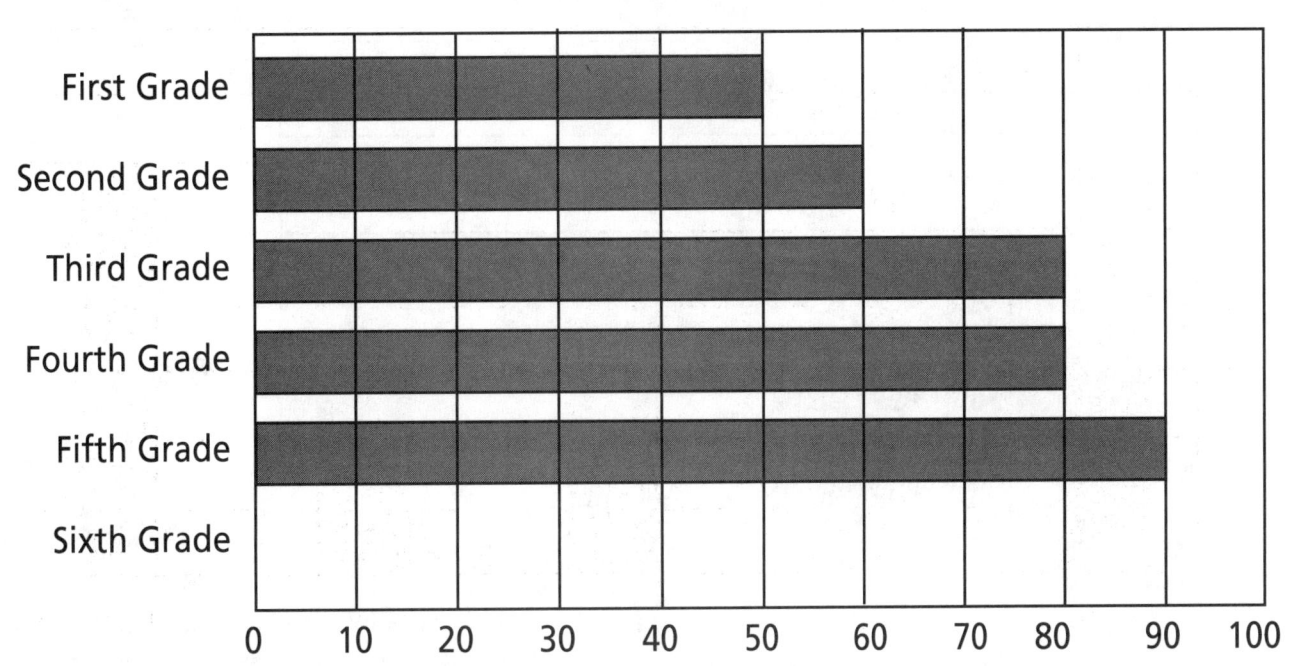

Grade	Cans
First Grade	50
Second Grade	60
Third Grade	80
Fourth Grade	70
Fifth Grade	90
Sixth Grade	0

0 10 20 30 40 50 60 70 80 90 100

A. Which grade collected the most cans? _____
How many cans were collected? _____

B. Which grade collected the fewest cans? _____
How many cans were collected? _____

C. How many more cans did the third grade students collect than the first grade students? _____

D. Which two grades collected the same amount of cans? _____

E. How many cans did the first, second, and third grade students collect in all? _____

F. The sixth grade students collected 20 more cans than the first grade students. How many did they collect? _____
Add this information on the graph.

Name_____

Crafty Items
Solving word problems/Picture graph

Mrs. Blair's Art Club held a craft sale to buy special art supplies. The picture graph shows how many items they sold.

Craft Items Sold

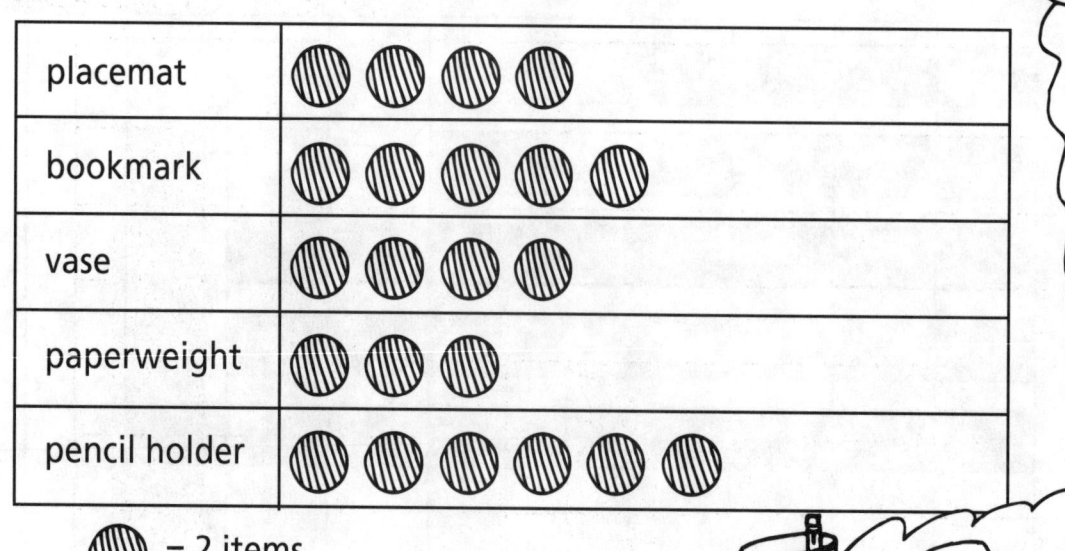

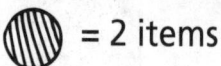

placemat	⊙ ⊙ ⊙ ⊙
bookmark	⊙ ⊙ ⊙ ⊙ ⊙
vase	⊙ ⊙ ⊙
paperweight	⊙ ⊙ ⊙
pencil holder	⊙ ⊙ ⊙ ⊙ ⊙

⊙ = 2 items

Use the picture graph to answer the questions.

A. How many placemats were sold? _____

B. How many more pencil holders were sold than vases? _____

C. Each vase cost $3.00. How much money was made from the sale of vases? _____

D. How many bookmarks were sold? _____
If each bookmark cost 25¢, how much money was made from their sale? _____

E. How many fewer paperweights were sold than bookmarks? _____

F. Each paperweight cost $2.00. How much money was made from the sale of paperweights? _____

© Frank Schaffer Publications, Inc.

Graphing Grade 3

Name_____

Museum Visitors
Solving word problems/Bar graph

The bar graph below shows how many people visited the Shoreline Maritime Museum last week. Use the graph to answer the questions.

Number of Visitors

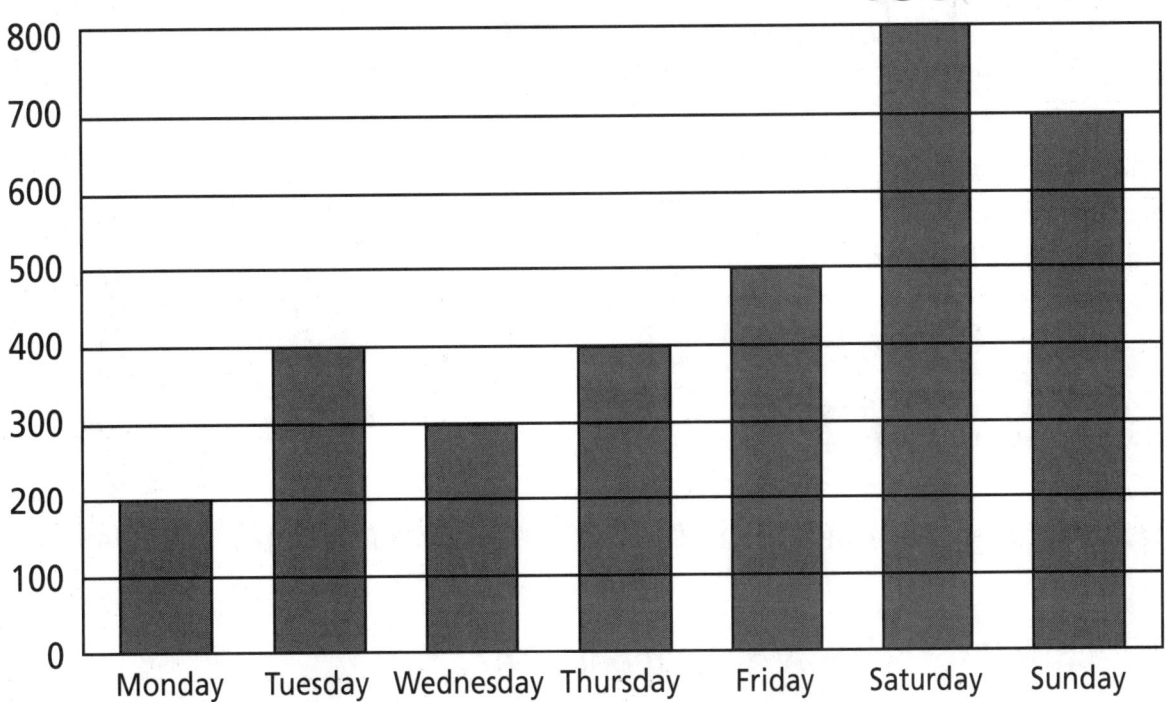

(Bar graph showing: Monday 200, Tuesday 400, Wednesday 300, Thursday 400, Friday 500, Saturday 800, Sunday 700)

A. How many people went to the museum on Monday? _____

B. How many people went to the museum on Tuesday? _____

C. Which two days had the same attendance? _____

D. Which two days had the lowest attendance? _____
What was the total number of visitors on those two days? _____

E. Look at the two days that had the highest attendance. What was the total number of visitors on those two days? _____

F. On which day were there four times as many visitors as on Monday?
_____

© Frank Schaffer Publications, Inc.

7

Graphing Grade 3

Name_____

Shopping Fun
Reading a circle graph

A circle graph is divided into sections like the slices of a pie. Look at the example below.

How Jeff Spent His Money

Jeff earned $6.00 helping with the family chores. Then he went to the toy store and spent his money on three items. The circle graph shows how much of his money went towards buying each item.

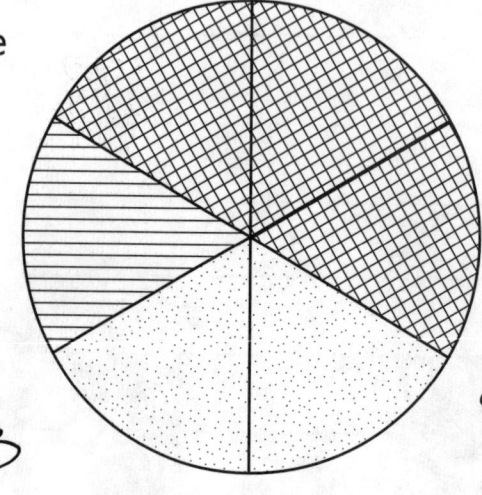

= toy car

= marbles

= comic book

Each section of the graph stands for $1.00. Use the graph to solve the problems.

A. What did Jeff spend most of his money on? _____

B. What did Jeff spend the least amount of money on? _____

C. Write how many dollars Jeff spent on the following toys:

　　toy car _____　　comic book _____　　marbles _____

D. How much of his money did Jeff spend on the car—$\frac{1}{2}$, $\frac{1}{3}$, or $\frac{1}{6}$? _____

E. How much of his money did Jeff spend on the comic book—$\frac{1}{2}$, $\frac{1}{3}$, or $\frac{1}{6}$? _____

F. How much of his money did Jeff spend on marbles—$\frac{1}{2}$, $\frac{1}{3}$, or $\frac{1}{6}$? _____

© Frank Schaffer Publications, Inc.

Graphing Grade 3

Name_____

All in the Family
Comparing parts of a whole/Circle graph

Emily has eight cats. The circle graph shows their various colorings.

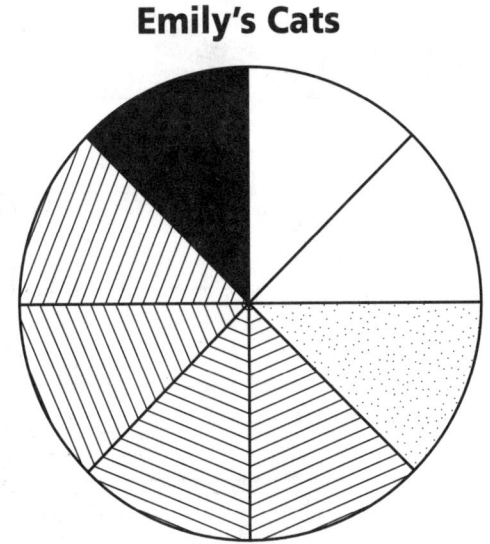

Emily's Cats

■ = black
□ = white
▒ = tan
▨ = striped

Answer the questions.

A. Are there more black cats or white cats? _____

B. How many striped cats are there? _____

C. Are there more tan cats or striped cats? _____

D. How many black cats and white cats are there altogether? _____

E. Which fraction describes how many cats are black— $\frac{1}{2}$, $\frac{1}{4}$, or $\frac{1}{8}$? _____

F. Which fraction describes how many cats are white— $\frac{1}{2}$, $\frac{1}{4}$, or $\frac{1}{8}$? _____

G. Which fraction describes how many cats are tan— $\frac{1}{2}$, $\frac{1}{4}$, or $\frac{1}{8}$? _____

H. Which fraction describes how many cats are striped— $\frac{1}{2}$, $\frac{1}{4}$, or $\frac{1}{8}$? _____

© Frank Schaffer Publications, Inc.

Graphing Grade 3

Name_____

How Many People? ·
Comparing populations/Picture graph

Use the picture graph to answer the questions.

Populations of Towns

Anytown	👤 👤 👤 👤
Sunnyplace	👤 👤 👤
Quietville	👤 👤
Someland	👤 👤 👤 👤
Happyglen	👤 👤 👤 👤 👤 👤 👤
Busyvale	👤 👤 👤 👤 👤 👤

👤 = 1,000 people

A. How many people live in Anytown? _____

B. How many people live in Happyglen? _____

C. Which town has a population of 3,000? _____

D. Which town has more people—Sunnyplace or Someland? _____

E. Which two towns have the same population? _____

F. How many more people live in Busyvale than in Quietville? _____

G. Which town has the fewest people? _____

H. Which town has the most people? _____

Name_____

Dinosaur Days
Comparing heights/Bar graph

The bar graph shows the heights of four dinosaurs.
Use the graph to answer the questions.

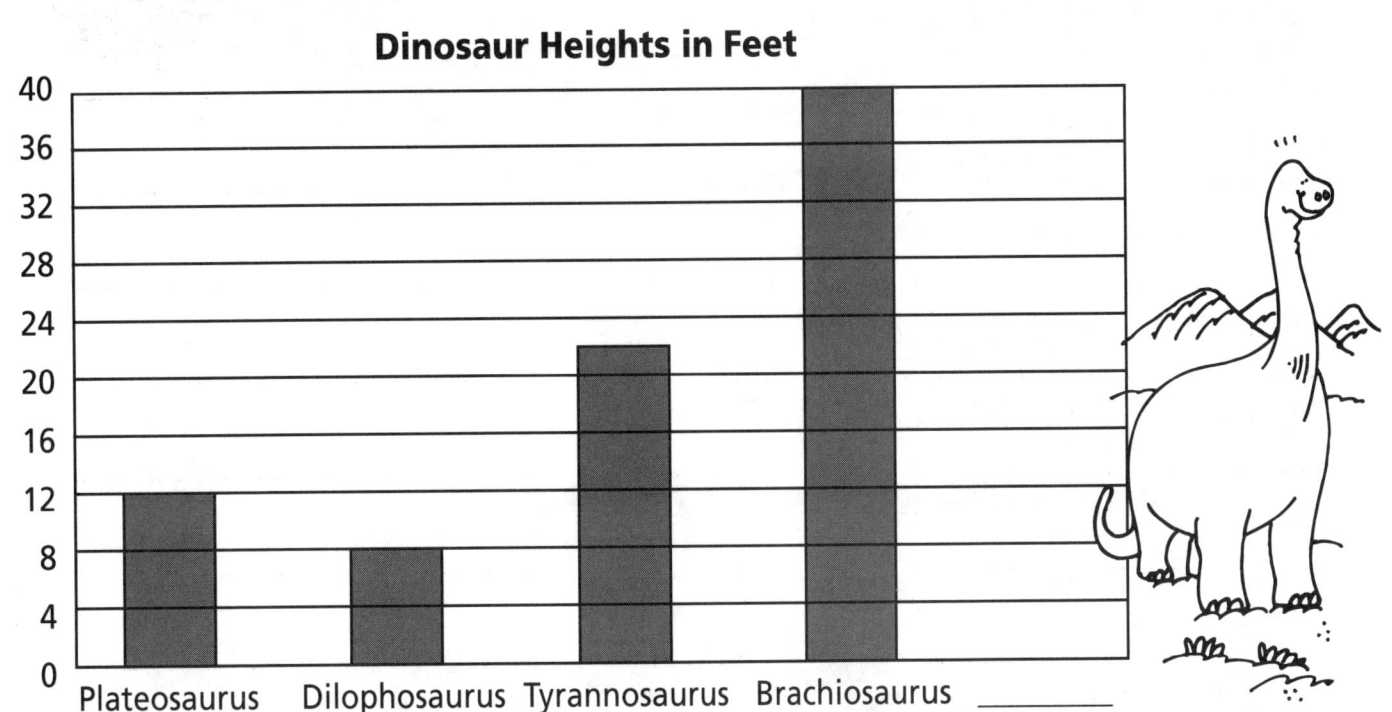

Dinosaur Heights in Feet

| Plateosaurus | Dilophosaurus | Tyrannosaurus | Brachiosaurus | _____ |

A. Which of the four dinosaurs was the tallest? Circle your answer.
   Plateosaurus   Dilophosaurus   Tyrannosaurus   Brachiosaurus

B. Which of the four dinosaurs was the shortest? Circle your answer.
   Plateosaurus   Dilophosaurus   Tyrannosaurus   Brachiosaurus

C. Which dinosaurs were less than 15 feet tall? Circle your answers.
   Plateosaurus   Dilophosaurus   Tyrannosaurus   Brachiosaurus

D. Which dinosaur was more than 15 feet tall, but shorter than 30 feet tall? Circle your answer.
   Plateosaurus   Dilophosaurus   Tyrannosaurus   Brachiosaurus

E. About how tall are you? _____
   Write your name on the graph and add a bar showing your height.

© Frank Schaffer Publications, Inc.

Graphing Grade 3

Long and Slippery

Comparing lengths/Bar graph

Ed, Ian, Joy, Lee, and Kim have garter snakes.
The bar graph shows the length of each snake.

Snake Lengths in Inches

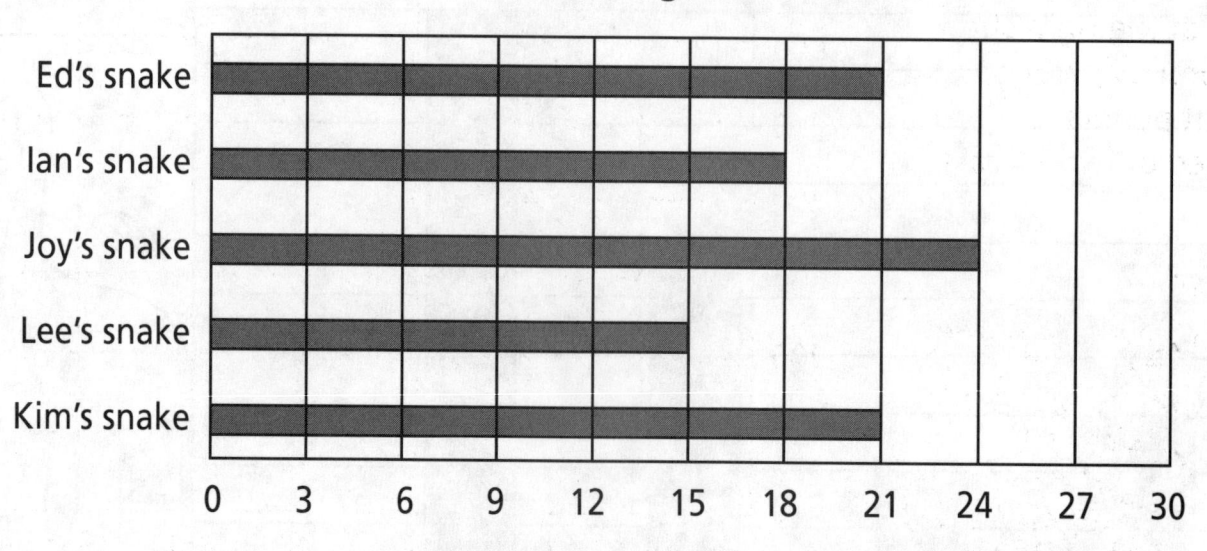

Snake	Length
Ed's snake	21
Ian's snake	18
Joy's snake	24
Lee's snake	15
Kim's snake	21

0 3 6 9 12 15 18 21 24 27 30

Answer the questions.

A. Which snake is the longest? _____ How long is it? _____

B. Which is the shortest snake? _____ How long is it? _____

C. Which snakes are the same length? _____
How long are they? _____

D. How many inches longer is Joy's snake than Ed's snake? _____

E. How many inches shorter is Lee's snake than Kim's? _____

F. How many snakes are longer than 20 inches? _____

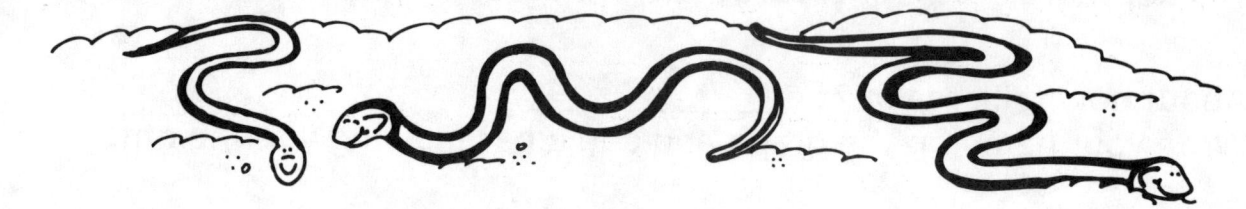

Name_____

Changes Over Time
Reading a line graph

A line graph can show how something changes over time. The dots on the graph at the right show how Grandview School's student population changed over the years.

Number of Students at Grandview School

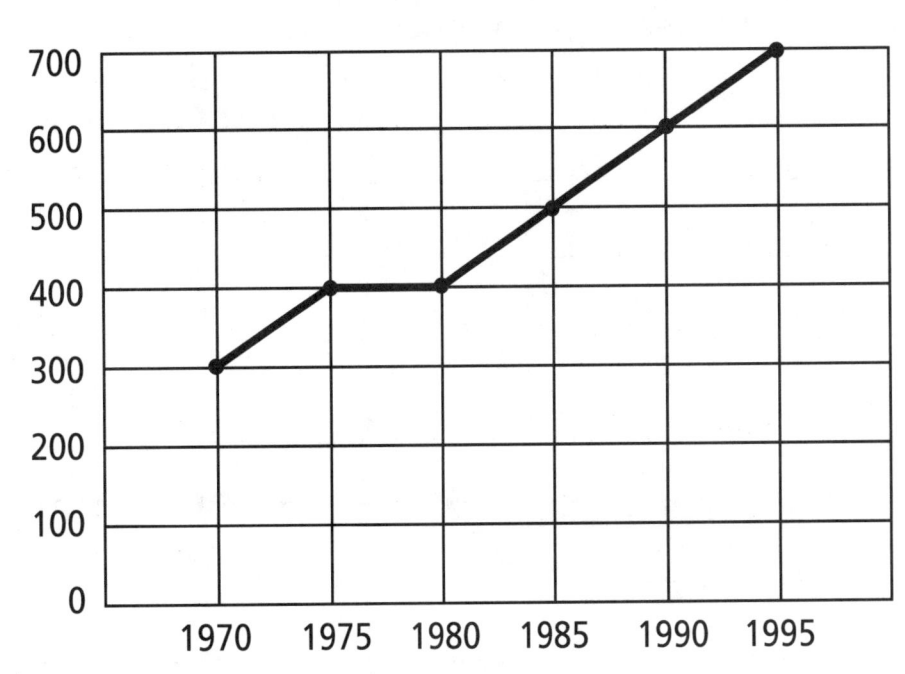

Look at the line graph and answer the questions.

A. How many students attended Grandview School in 1970? _____

B. How many students attended the school in 1980? _____

C. When did the student population stay the same? _____

D. How did the population change between 1985 and 1995?
_____

E. How many more students attended the school in 1995 than in 1970?
_____

F. Do you think the school population will go up or go down by the year 2000? Why? _____
_____

Name_____

Lemonade Sales

Solving word problems/Line graph

Curtis and Tanya set up a lemonade stand. The line graph shows how many glasses they sold in one week. Use the graph to answer the questions.

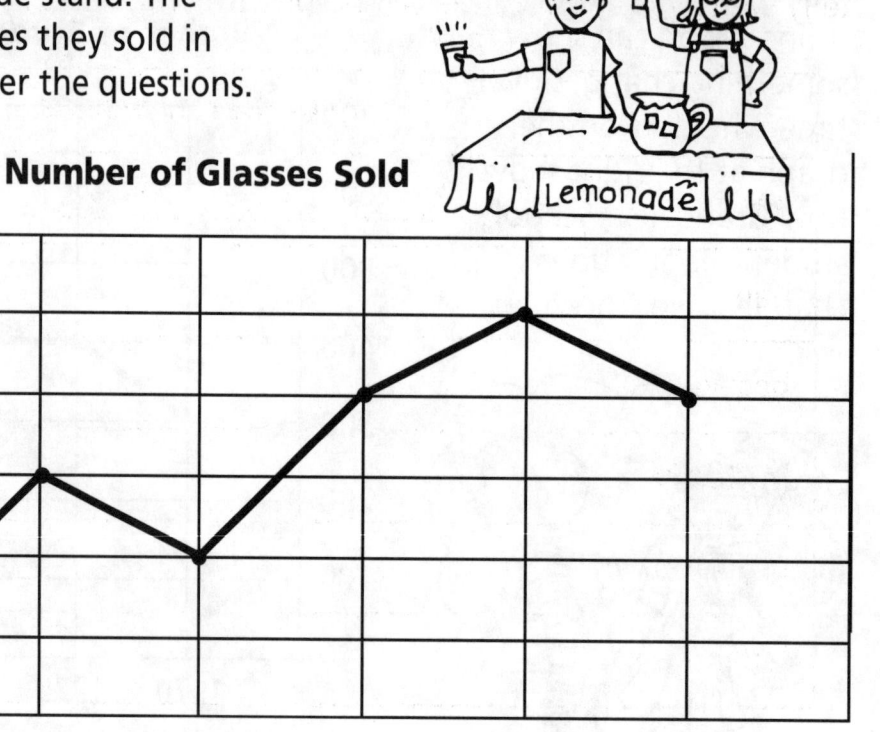

Number of Glasses Sold

A. How many glasses were sold on Monday? _____

B. How many glasses were sold on Friday? _____

C. How many more glasses were sold on Wednesday than on Tuesday? _____

D. Which day had the fewest sales? _____

E. Which day had the most sales? _____

F. If each glass of lemonade cost 50¢, how much money did Curtis and Tanya make on Thursday? _____

G. If each glass of lemonade cost 50¢, what is the most money that the children made in one day? _____

Name_____

An Insect Hunt
Making computations/Completing a picture graph

Kelly went hunting for insects. The picture graph lists the different insects she found.

	Number of Insects
ant	△ △ △ △ △ △ △
bee	
beetle	
butterfly	
grasshopper	
ladybug	

△ = 2 insects

Read the clues. Answer the questions. Then fill in the graph.

A. There were 8 more ants than bees. How many bees were there? _____

B. There were twice as many beetles than bees. How many beetles were there? _____

C. There were half as many butterflies as there were beetles. How many butterflies were there? _____

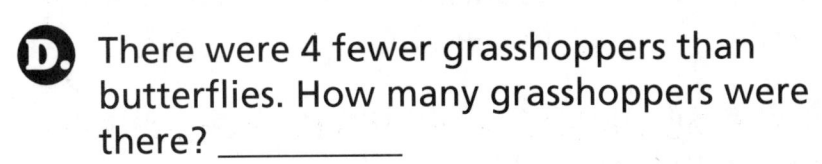

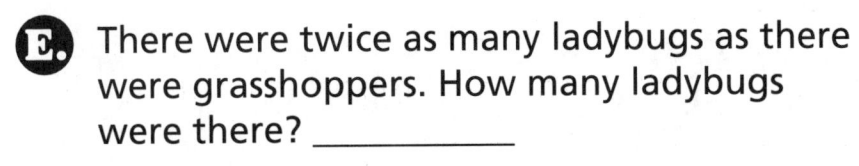

D. There were 4 fewer grasshoppers than butterflies. How many grasshoppers were there? _____

E. There were twice as many ladybugs as there were grasshoppers. How many ladybugs were there? _____

© Frank Schaffer Publications, Inc.

Graphing Grade 3

Name_____

Sports Day Fun
Making computations/Completing a bar graph

The third grade students held a Sports Day at school. The bar graph lists the different events they had and the number of children in each race.

Number of Children

crabwalk race										
obstacle course	■	■	■	■	■	■				
running race										
relay race										
sack race										
three-legged race										

0   5   10   15   20   25   30   35   40   45

Read the clues. Answer the questions. Then fill in the graph.

A. There were only half as many students in the crabwalk race as there were in the obstacle course. How many children competed in the crabwalk race?
_____

B. The obstacle course and the relay race had the same number of students. How many children were in the relay race? _____

C. There were 10 more children competing in the sack race than there were in the crabwalk race. How many children were in the sack race? _____

D. There were 5 fewer children in the running race than there were in the sack race. How many children were in the running race? _____

E. There were twice as many students in the three-legged race as there were in the running race. How many children were in the three-legged race?
_____

© Frank Schaffer Publications, Inc.

16

Graphing Grade 3

Name_____

A Stamp Collection
Logical thinking/Completing a picture graph

Shawn collects stamps from around the world. Read the clues to figure out how many stamps Shawn has collected from Canada, Mexico, France, and Brazil. Write your answers on the chart.

Clues:
- ◆ There are 3 times as many stamps from Mexico as there are from France.
- ◆ There are 8 fewer stamps from Canada than from Mexico.
- ◆ There are 4 more stamps from France than from Brazil.
- ◆ There are 4 stamps from Brazil.

Name of Country	Number of Stamps
Canada	
Mexico	
France	
Brazil	

Now use the information from your chart to create a picture graph.

Shawn's Stamp Collection

Canada	
Mexico	
France	
Brazil	

☐ = 4 stamps

Name_____

T-Shirt Count
Logical thinking/Completing a circle graph

Jan grouped her 12 T-shirts according to these colors: red, blue, green, and yellow. Read the clues and figure out how many T-shirts Jan has of each color. Write your answers on the chart.

Clues:

♦ There are twice as many blue T-shirts as there are red T-shirts.

♦ There is the same number of green T-shirts and yellow T-shirts.

♦ There is 1 more blue T-shirt than there are green T-shirts.

♦ There are 2 red T-shirts.

T-Shirt Color	Number of T-Shirts
red	
blue	
green	
yellow	

Now use the information from your chart to create a circle graph. First color the key so that the rectangles match their colors. Then color each section of the circle to show how many T-shirts Jan has of each color.

Jan's T-Shirts

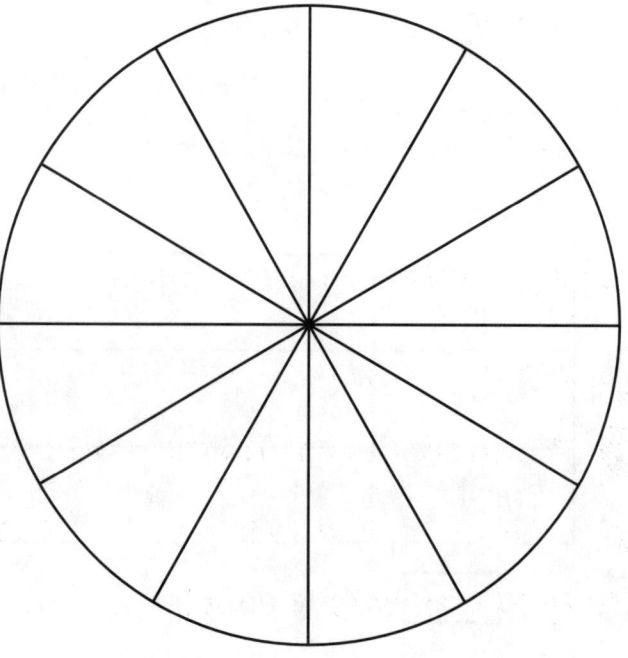

☐ = red   ☐ = green
☐ = blue ☐ = yellow

© Frank Schaffer Publications, Inc.

Graphing Grade 3

Name_____

Tasty Sandwiches
Using data from a chart to a make picture graph

Karly took a survey to see what kinds of sandwiches people like the best. She wrote the information on a chart. Use the chart to complete the picture graph below.

Sandwich	Number of People
cheese	6
ham	10
tuna	2
chicken	4
peanut butter	8

Favorite Sandwiches

cheese	
ham	
tuna	
chicken	
peanut butter	

◯ = 2 people

Look at the graph and answer the questions.

A. Which sandwich was the most popular? _____

B. Which sandwich was the least popular? _____

C. How many more people liked cheese than chicken? _____

D. How many more people liked peanut butter than tuna? _____

© Frank Schaffer Publications, Inc.

Graphing Grade 3

Name_____

How Speedy?

Using data from a chart to make a bar graph

The chart at the right compares how fast some land animals run. Use the chart to help you complete the bar graph below.

Animal	Miles Per Hour
cheetah	70
elephant	25
gazelle	50
horse	45
house cat	30
human	20

Animal Speeds

cheetah														
elephant														
gazelle														
horse														
house cat														
human														

0   5   10   15   20   25   30   35   40   45   50   55   60   65   70
Miles Per Hour

Look at the graph and answer the questions.

A. Which land animal is the fastest? _____

B. Which animal is faster—a gazelle or an elephant? _____

C. Could a person outrun a cat? _____

D. Could a cheetah outrun a horse? _____

© Frank Schaffer Publications, Inc.   20   Graphing Grade 3

Name_____

All Kinds of Books

Using data from a chart to make a circle graph

Type of Book	Number of Books
fiction	3
science	2
biography	2
poetry	1

Lisa read 8 library books last month. The chart lists the types of books she read. It also shows how many of each kind of book she read.

Use the chart to create a circle graph. First color the key. Make each rectangle a different color. Then color each section of the circle graph to match the information on the chart.

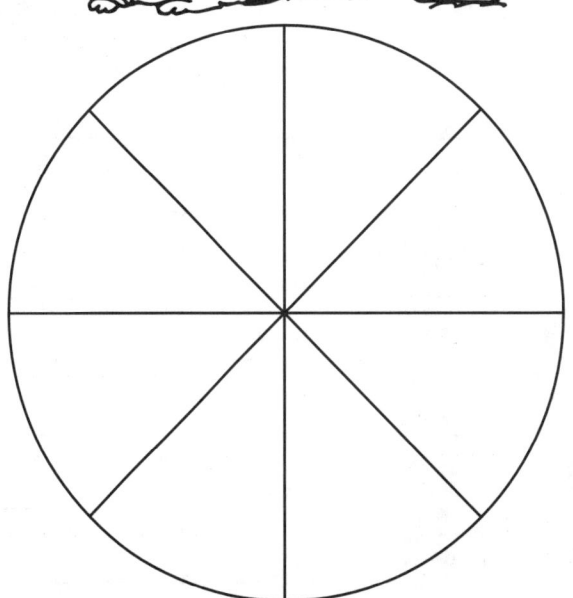

☐ = fiction   ☐ = biography
☐ = science   ☐ = poetry

Look at the graph and answer the questions.

A. What type of book did Lisa read the most? _____

B. How many more science books did she read than poetry books? _____

C. Which fraction describes how many of the books were biographies— $\frac{1}{8}$ or $\frac{2}{8}$? _____

D. Which fraction describes how many of the books were science books— $\frac{1}{4}$ or $\frac{1}{8}$? _____

E. Which fraction describes how many of the books were fiction— $\frac{3}{4}$ or $\frac{3}{8}$? _____

© Frank Schaffer Publications, Inc.

Graphing Grade 3

Name_____

Oletown's Rain...........

Using data from a chart to make a bar graph

The chart below shows the amount of rain that Oletown received in one year.

Month	Jan.	Feb.	Mar.	Apr.	May	June	July	Aug.	Sept.	Oct.	Nov.	Dec.
Rainfall in Inches	6	5	5	4	3	2	2	1	3	4	7	5

Use the information on the chart to complete the bar graph.

Annual Rainfall in Oletown

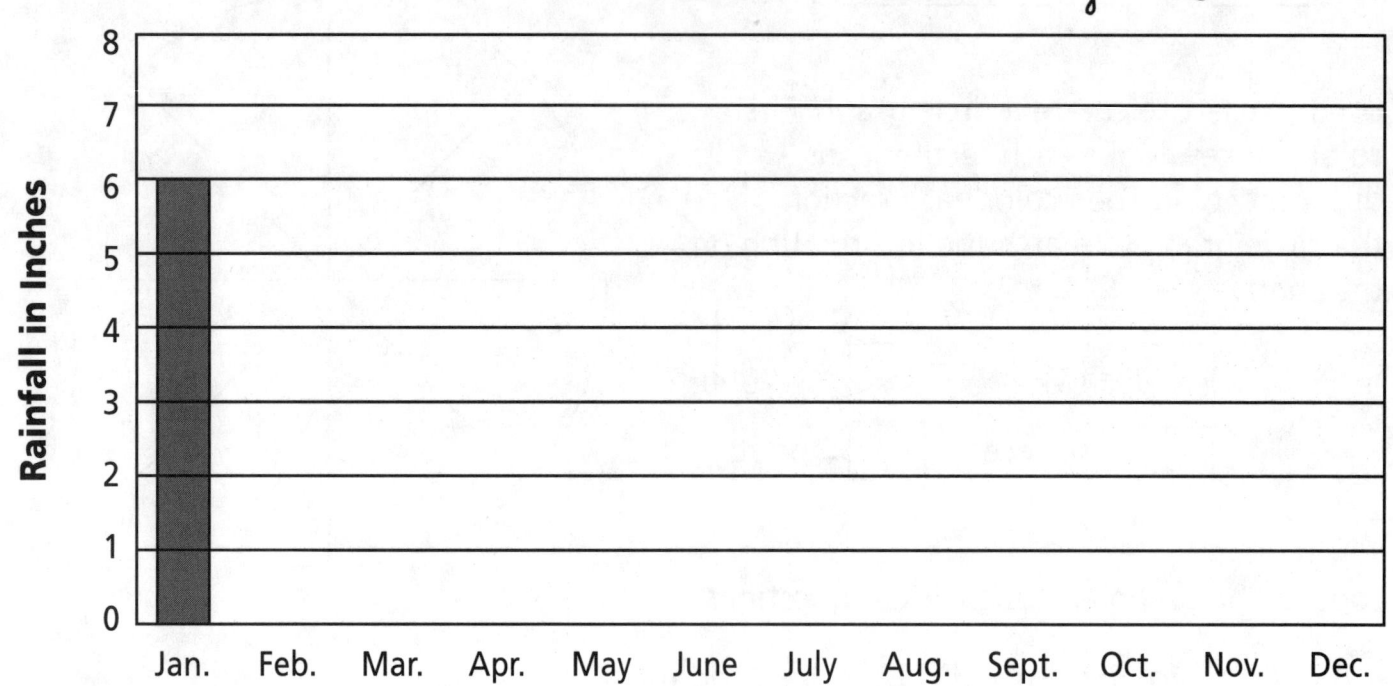

Answer the questions.

A. Which month received the most rain? _____

B. Which month received the least amount of rain? _____

C. Which months had more than 5 inches of rain? _____

D. Which months had less than 3 inches of rain? _____

Name_____

Aquarium Animals
Reading tally marks/Creating a bar graph

Mrs. Turner's students went to the aquarium. They used a tally to keep track of the different animals. A tally mark looks like a stick. One mark stands for one animal. The fifth tally mark crosses the other four, so 卌 stands for five animals. The chart shows what that the students saw.

angelfish	crab	dolphin
卌 卌 ||	卌 ||	卌
eel	shark	stingray
卌 卌		||||	||

Use the chart to make a bar graph.

Number of Sea Creatures Seen at the Aquarium

angelfish													
crab													
dolphin													
eel													
shark													
stingray													

0 1 2 3 4 5 6 7 8 9 10 11 12 13

Answer the questions.

A. How many sharks are there? _____

B. How many eels are there? _____

C. Are there more crabs or dolphins? _____

D. How many more angelfish are there than stingrays? _____

Name_____

A Choice of Graphs
Displaying information on two different graphs

Sometimes you can use different graphs to give the same information. For example, look at the picture graph. It shows how many boxes of Crunch Cereal were sold at five grocery stores last Saturday.

Boxes of Crunch Cereal Sold

Smart Market	■ ■ ■ ■
OK Food Mart	■ ■ ■
Super Foods	■ ■ ■ ■ ■
Grand Groceries	■ ■ ■ ■ ■
Fresh Picks	■ ■

■ = 5 boxes

Now use the picture graph to make a bar graph showing the same information.

```
25
20 ████
15 ████
10 ████
 5 ████
 0 ████
   Smart Market   OK Food Mart   Super Foods   Grand Groceries   Fresh Picks
```

Look at both graphs. Answer the questions.

A. Do both graphs help you compare how many boxes were sold at the different stores? _____

B. Which graph do you prefer? _____ Why? _____

_____

_____

© Frank Schaffer Publications, Inc.

Graphing Grade 3

Name_____

Carnival Time
Gathering data from two graphs

Brad and Meg went to the carnival. They each had $10.00 to spend. The circle graphs show how they spent their money.

How Brad Spent His Money

How Meg Spent Her Money

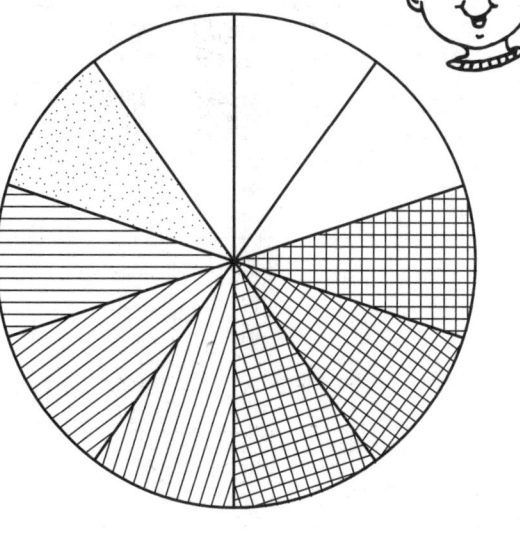

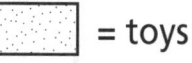

☐ = rides
▦ = games
▥ = food
⋯ = toys

Look at both graphs to help you answer the questions.

A. Did Brad spend more money on rides or on toys? _____

B. Did Meg spend more money on rides or on toys? _____

C. Who spent more money on games? _____

D. How much money did Brad spend on food? _____

E. How much money did Meg spend on food? _____

F. What did Brad and Meg spend the same amount of money on? _____

How much money did each child spend? _____

© Frank Schaffer Publications, Inc.

25

Graphing Grade 3

Name_____

A Penny Drive
Gathering data from two graphs

The Boys Scouts and Girl Scouts held a penny drive to help needy families. The bar graphs show how many pennies they collected over a period of five days.

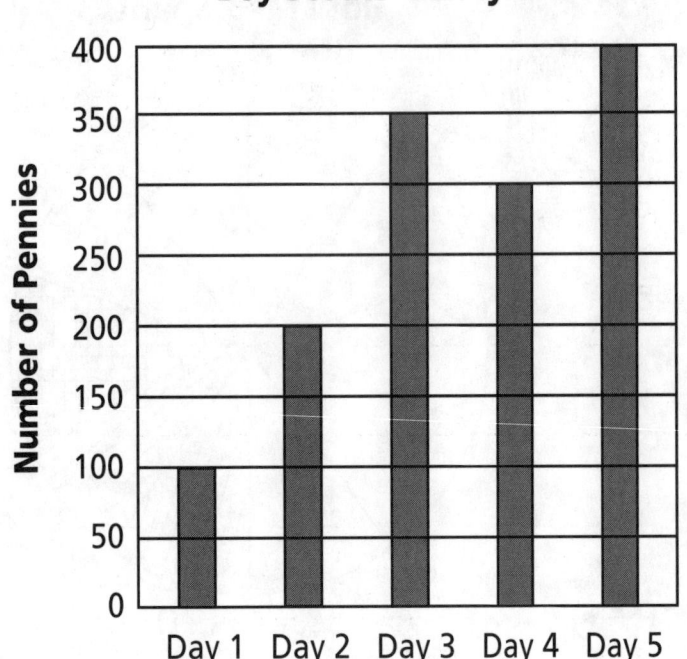

Boy Scouts' Penny Drive

Girl Scouts' Penny Drive

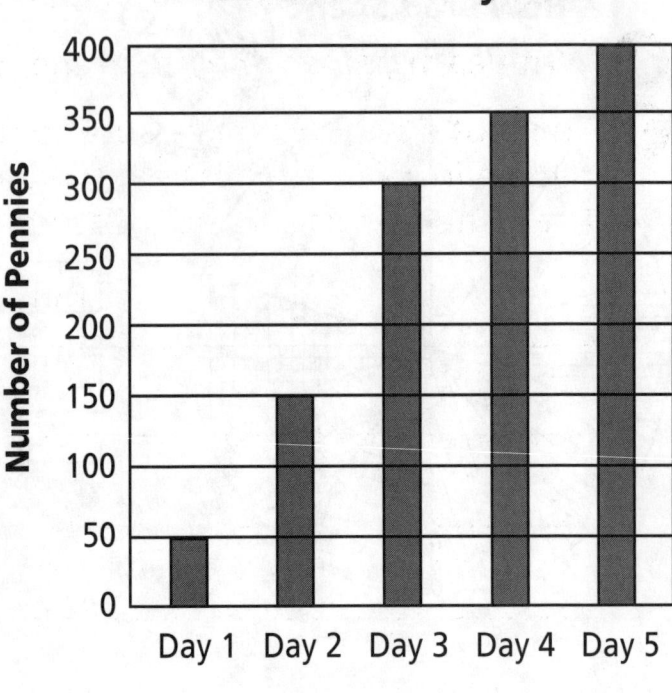

Look at both graphs to help you answer the questions.

A. How many pennies did the Boy Scouts collect on Day 1? _____
How many pennies did the Girl Scouts collect on the same day? _____

B. How many pennies did the Boy Scouts collect in all during the first three days? _____ How many pennies did the Girl Scouts collect during the first three days? _____

C. On which day did the Boy Scouts and Girl Scouts collect the same amount?
_____

D. Who collected more pennies on Day 4? _____

E. How many pennies did each group collect in all at the end of five days?
Boy Scouts _____   Girl Scouts _____

Name_____

Sizes of Families
Collecting data/Creating a picture graph

Make a picture graph showing the sizes of some families.

1. Ask five people how many members are in their families. Record their answers on a sheet of paper.

2. Write each person's name on the graph below.

3. Think of a symbol you can use to show the number of people in each family. Draw the symbol below the graph. Draw the correct number of symbols for each family.

Family Sizes

= 1 person

Look at the completed graph. Answer the questions.

A. Who had the largest family? _____
How many people are in that family? _____

B. Who had the smallest family? _____
How many people are in that family? _____

C. Suppose you wanted to show how many males and how many females were in each family? What changes could you make on your graph to do that? _____

_____

© Frank Schaffer Publications, Inc.

Graphing Grade 3

Name_____

Off to School......................
Collecting data/Creating a bar graph

Make a picture graph showing how some students get to school.

1. Ask 12 students how they get to school (walk, ride a bike, ride in a car, or go by bus). Record their answers on a sheet of paper.

2. Count the number in each category.

3. Make bars on the graph below to match your findings.

How Students Get to School

walk													
bike													
car													
bus													

0 1 2 3 4 5 6 7 8 9 10 11 12

Number of Students

Look at the completed graph. Answer the questions.

A. How did most of the students get to school? _____

B. Did more students go by car or by bus? _____

C. Did more students walk or ride a bike? _____

D. How do you get to school? _____
Add that information to the graph.

Name_____

Coin Throw
Collecting data/Creating a circle graph

Make a circle graph showing how many times a coin lands with heads facing up or with tails facing up.

How a Coin Landed When Tossed

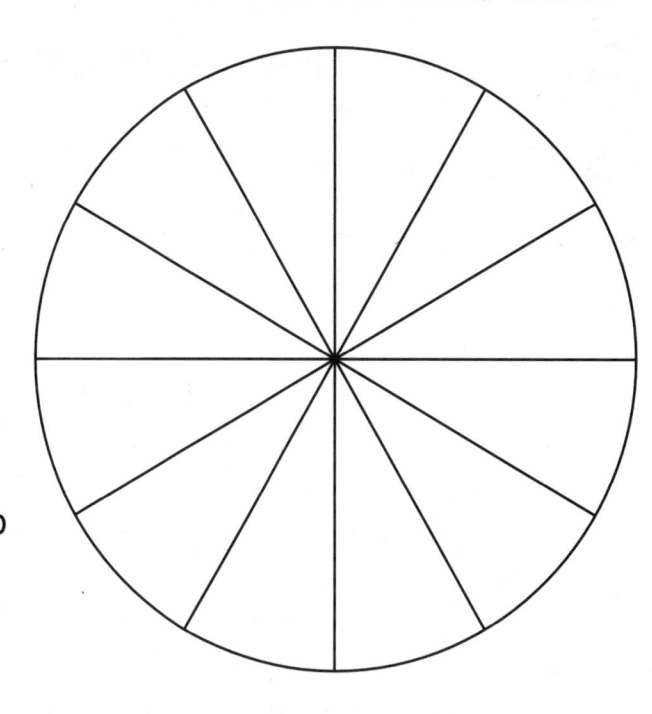

1. Toss a coin 12 times. Each time record whether it lands with heads facing up or with tails facing up.

2. Color the rectangles in the key below. Color the circle graph to match your findings.

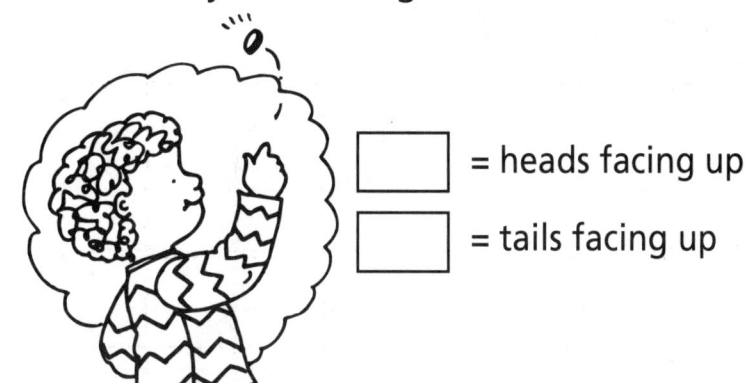

☐ = heads facing up

☐ = tails facing up

Look at the completed graph. Answer the questions.

A. How many times did the coin land with heads facing up? _____
What fraction could you use to show how much of the time it did this?
_____

B. How many times did the coin land with tails facing up? _____
What fraction could you use to show how much of the time it did this?
_____

C. Compare your graph with that of another student. Were the graphs similar or different? Explain your answer.
_____
_____

© Frank Schaffer Publications, Inc.

Graphing Grade 3

Name_____

TV Watch
Collecting data/Creating a line graph

Make a line graph showing how many hours of TV you watched in one week.

1. Keep track of how many hours of TV you watch each day for a week. Write the hours on a sheet of paper.
2. Record your findings on the graph below. For each day, draw a dot where the lines for the number of hours and the day meet.
3. Connect the dots with a ruler.

Number of Hours I Watched TV

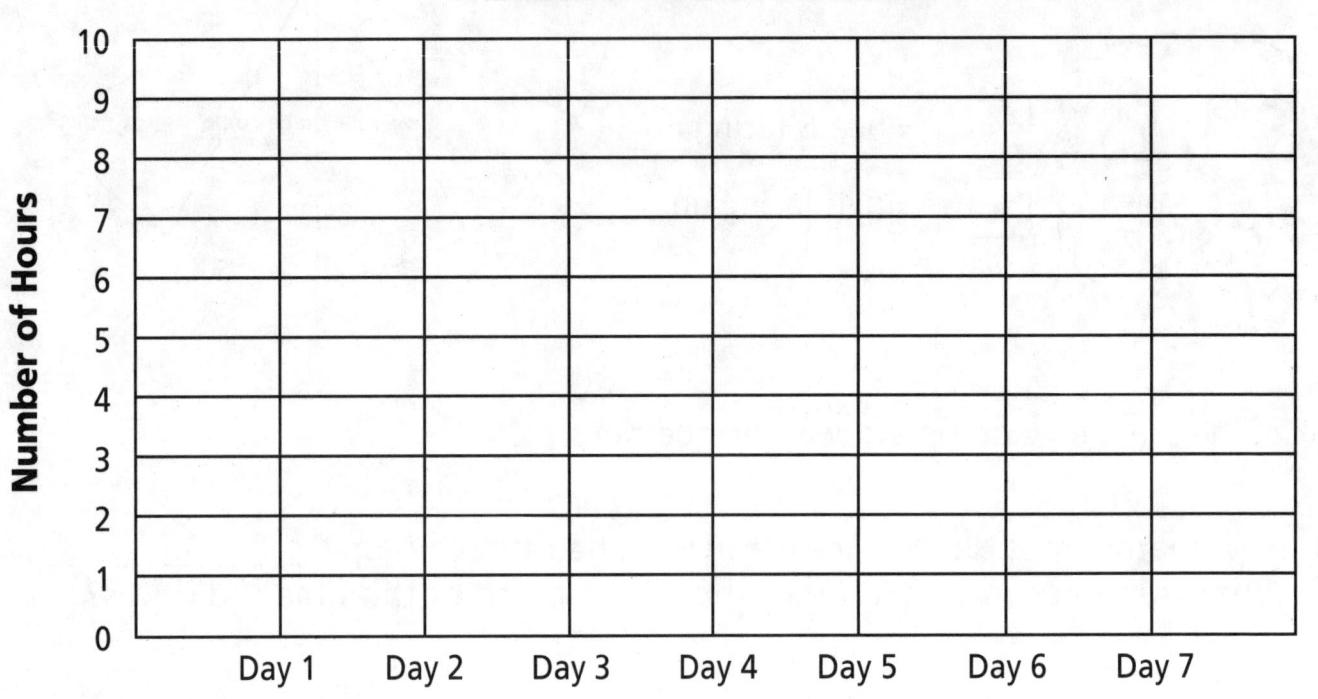

Number of Hours	Day 1	Day 2	Day 3	Day 4	Day 5	Day 6	Day 7
10							
9							
8							
7							
6							
5							
4							
3							
2							
1							
0							

Look at the completed graph. Answer the questions.

A. On which day did you spend the most time watching TV? _____
B. When did you spend the least amount of time watching TV? _____
C. How many hours did you spend watching TV in one week? _____

ANSWERS

Page 3
A. 7
B. acrobats
C. 3
D. 3
E. clowns
F. animal trainers
G. 25

Page 4
A. 4
B. 7
C. 3
D. apples
E. oranges
F. watermelons
G. apples
H. 26

Page 5
A. 5th, 90
B. 1st, 50
C. 30
D. 3rd, 4th
E. 190
F. 70
Student should draw a bar on the graph to show that 70 cans were collected by the 6th grade students.

Page 6
A. 8
B. 4
C. $24.00
D. 10, $2.50
E. 4
F. $12.00

Page 7
A. 200
B. 400
C. Tuesday, Thursday
D. Monday, Wednesday; 500
E. 1,500
F. Saturday

Page 8
A. a toy car
B. comic book
C. car—$3.00, comic book—$1.00, marbles—$2.00
D. 1/2
E. 1/6
F. 1/3

Page 9
A. white cats
B. 4
C. striped
D. 3
E. 1/8
F. 1/4
G. 1/8
H. 1/2

Page 10
A. 4,000
B. 7,000
C. Sunnyplace
D. Someland
E. Anytown, Someland
F. 4,000
G. Quietville
H. Happyglen

Page 11
A. Brachiosaurus
B. Dilophosaurus
C. Plateosaurus, Dilophosaurus
D. Tyrannosaurus
E. Answer will vary.

Page 12
A. Joy's snake, 24 inches
B. Lee's snake, 15 inches
C. Ed's and Kim's snakes, 21 inches
D. 3 inches
E. 6 inches
F. 3 snakes

Page 13
A. 300
B. 400
C. between 1975 and 1980
D. The population rose by 200.
E. 400
F. Answer will vary.

Page 14
A. 10
B. 20
C. 10
D. Tuesday
E. Saturday
F. $5.00
G. $12.50

Page 15
A. 6
B. 12
C. 6
D. 2
E. 4
Graph should be completed as shown:

Number of Insects	
ant	△△△△△△
bee	△△△
beetle	△△△△△△
butterfly	△△△
grasshopper	△
ladybug	△△

Page 16
A. 15
B. 30
C. 25
D. 20
E. 40
Graph should be completed as shown:

Page 17
Canada – 16 stamps
Mexico – 24 stamps
France – 8 stamps
Brazil – 4 stamps
Graph should be completed as shown:

Page 18
red – 2 t-shirts
blue – 4 t-shirts
green – 3 t-shirts
yellow – 3 t-shirts
Graph should be colored to show how many t-shirts there are of each color

ANSWERS

Page 19
Graph should be completed as shown:

Favorite Sandwiches

cheese	○ ○ ○
ham	○ ○ ○ ○ ○
tuna	○
chicken	○ ○
peanut butter	○ ○ ○ ○

○ = 2 people

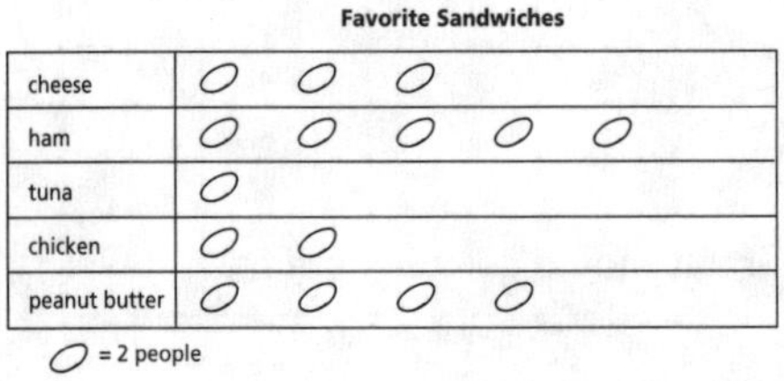

A. ham
B. tuna
C. 2
D. 6

Page 20
Graph should be completed as shown:

Animal Speeds

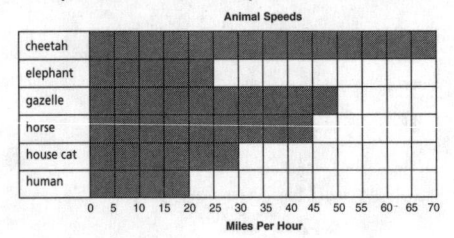

A. cheetah
B. gazelle
C. no
D. yes

Page 21
Graph should be colored to show that there are 3 fiction books, 2 science books, 2 biography books, and 1 poetry book.

A. fiction
B. 1
C. 2/8
D. 1/4
E. 3/8

Page 22
Graph should be completed as shown:

Annual Rainfall in Oletown

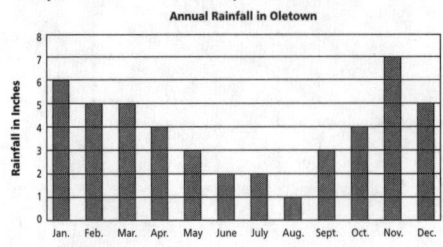

A. November
B. August
C. January, November
D. June, July, August

Page 23
Graph should be completed as shown:

Number of Sea Creatures Seen at the Aquarium

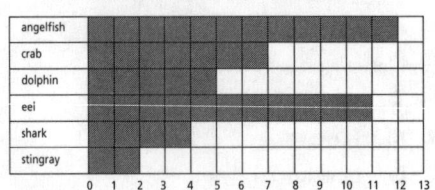

A. 4
B. 11
C. crabs
D. 10

Page 24
Graph should be completed as shown:

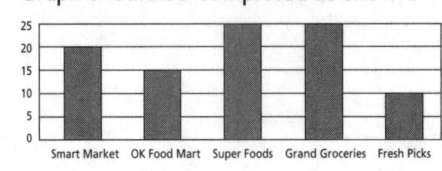

A. yes
B. Answers will vary.

Page 25
A. rides
B. toys
C. Brad
D. $3.00
E. $2.00
F. rides, $3.00

Page 26
A. Boy Scouts – 100
   Girl Scouts – 50
B. Boy Scouts – 650
   Girl Scouts – 500
C. Day 5
D. Girl Scouts
E. Boy Scouts – 1,350
   Girl Scouts – 1,250

Page 27
Graph will vary.
A. Answer will vary.
B. Answer will vary.
C. Make one symbol for male and one symbol for female. Use both symbols on the graph.

Page 28
Graph will vary.
A. Answer will vary.
B. Answer will vary.
C. Answer will vary.
D. Answer will vary.

Page 29
Graph will vary.
A. Answer will vary.
B. Answer will vary.
C. Answer will vary.

Page 30
Graph will vary.
A. Answer will vary.
B. Answer will vary.
C. Answer will vary.